T0135587

Modal logic

Compact companion 2

Herman Ruge Jervell
University of Oslo

Bibliographic information published by the Deutsche Nationalbibliothek

The Deutsche Nationalbibliothek lists this publication in the Deutsche
Nationalbibliografie; detailed bibliographic data are available
in the Internet at http://dnb.d-nb.de.

ISBN 978-3-8325-3302-1

Logos Verlag Berlin GmbH
Comeniushof, Gubener Str. 47,
10243 Berlin
Tel.: +49 (0)30 42 85 10 90
Fax: +49 (0)30 42 85 10 92
INTERNET: http://www.logos-verlag.de

Contents

Preface

In modal logic we treat the notion of multiple forms of truth. In practical science we use this notion. It may be that we can reduce statements in biology to statements of physics, but in practice we rarely do. We work with two — hopefully interrelated — notions of truth. And then we come to the realm of modal logic. Traditionally there are many sources for modal logic:

Temporality We distinguish between what is true in past, present or future.

Knowledge We distinguish between what just happens to be true, and what we know as true — perhaps as given by a formal proof.

Computations We distinguish between the intensional description of a computation as given by a code, and the extensional description as given by an input/output relation.

Agents We may have many agents and communication between them generating different perspectives.

Modal logic gives a frame work for arguing about these distinctions. Some of the high points are

Temporality The possible world semantics as given by Stig Kanger and Saul Kripke connects formal systems for modal logic and geometrical assumptions about the temporal relation.

v

Knowledge The Gödel incompleteness theorem makes clear the gap between what we can prove and what is true in formal systems strong enough to simulate syntactical operations. The completeness theorem of Robert Solovay shows that a decidable modal logic can capture the usual way of expressing incompleteness.

Computations Michael Rabin gave a decidable theory for the behavior of finite state automata with infinite streams as input. This is one of the most interesting decidable theories and has led to a number of ways of capturing the intensional/extensional distinction. We have used the analysis of Sergei Vorobyov of determinacy of games on finite arenas as an access to this very interesting theory.

Agents The problem of getting consensus between agents using narrow communication channels. Here we have used the the problem of the Byzantine generals by Leslie Lamport as an introduction and have included the stochastic solution by Michael Rabin using a global coin toss.

As background for the theory here we refer to the first compact companion — "Logic and Computations". But any reasonable textbook in logic should do.

Frege systems

1

1.1 Language of modal logic

Modal logic is an extension of propositional logic using two new unary connectives

- $\Box$ — necessary — called box

- $\Diamond$ — possibly — called diamond

We work with classical logic and the two new connectives satisfy de Morgans laws

$$\Box F \Leftrightarrow \neg \Diamond \neg F$$
$$\Diamond F \Leftrightarrow \neg \Box \neg F$$

We shall not consider extensions to predicate logic.

1.2 The calculi of modal logic

In the usual Frege version of classical propositional logic we have the rules modus ponens and the three axiom schemes **S**, **K** and **N**:

Modus ponens: $\vdash F$ and $\vdash F \to G \Rightarrow \vdash G$

S: $\vdash (A \to (B \to C)) \to ((A \to B) \to (A \to C))$

K: $\vdash A \to (B \to A)$

N: $\vdash ((A \to \bot) \to \bot) \to A$

The extension of the language to conjunction, disjunction and negation is simple and is left to the reader. To get propositional modal logic we add one rule — necessitation — and some new axioms which are often picked from the list below:

Necessitation: $\vdash F \Rightarrow \vdash \Box F$

Normal: $\vdash \Box(F \to G) \to (\Box F \to \Box G)$

T: $\vdash \Box F \to F$

B: $\vdash F \to \Box \Diamond F$

4: $\vdash \Box F \to \Box \Box F$

5: $\vdash \Diamond F \to \Box \Diamond F$

P: $\vdash F \to \Box F$

Q: $\vdash \Diamond F \to \Box F$

R: $\vdash \Box \Box F \to \Box F$

W: $\vdash \Box(\Box F \to F) \to \Box F$

M: $\vdash \Box \Diamond F \to \Diamond \Box F$

N: $\vdash \Diamond \Box F \to \Box \Diamond F$

D: $\vdash \Box F \to \Diamond F$

L: $\vdash \Box(F \wedge \Box F \to G) \vee \Box(G \wedge \Box G \to F)$

All the system we shall consider contains **Necessitation** and **Normal**. They are called normal modal systems. Some of the noteworthy systems are

Kripke: Least normal system — propositional logic + necessitation + normal

T: Kripke + axiom T

K4: Kripke + axiom 4

S4: Kripke + axiom T + axiom 4

S5: Kripke + axiom T + axiom 5

Gödel-Löb: Kripke + axiom W

1.3 Kripke models

A Kripke model over a language is given by

Frame: A set of points $\mathcal{K}$ and a binary relation between them $\prec$

Truthvalues: To each point in the frame $\mathcal{K}$ is assigned a truthvalue to the atomic formulas. The truth value is extended to arbitrary formula by

$$\alpha \models \neg F \Leftrightarrow \text{ not } \alpha \models F$$
$$\alpha \models F \wedge G \Leftrightarrow \alpha \models F \text{ and } \alpha \models G$$
$$\alpha \models \Box F \Leftrightarrow \forall \beta \succ \alpha \ . \ \beta \models F$$
$$\alpha \models \Diamond F \Leftrightarrow \exists \beta \succ \alpha \ . \ \beta \models F$$

Similarly for formulas built up from other connectives.

1.4 Completeness

The usual completeness theorem in propositional logic — using maximal consistent sets of formulas — can be generalized to modal logic.

In propositional logic we get completeness by the following steps

- A consistent set of formulas has a model

  - Every consistent set of formulas can be extended to a maximal consistent set
  - A maximal consistent set of formulas give an assignment of truth values making all formulas in the set true

The main work is done by extending a consistent set to a maximal consistent set. This is done by enumerating all formulas F_i where $i < \omega$. If we then start with a consistent set Γ_0, we extend it to $\Gamma = \cup_i \Gamma_i$ by

$$\Gamma_{i+1} = \left\{ \begin{array}{ll} \Gamma_i \cup \{F_i\} & \text{, if it is consistent} \\ \Gamma_i \cup \{\neg F_i\} & \text{, otherwise} \end{array} \right.$$

In modal logic

- Every consistent set of formulas can be extended to a maximal consistent set

- We build a frame of all maximal consistent sets — the binary relation between two sets $\mathcal{U}$ and $\mathcal{V}$ is given by $\{F | \Box F \in \mathcal{U}\} \subseteq \mathcal{V}$

- A formula is true in $\mathcal{U}$ if it is contained in it — this is constructed so for atomic formulas and proved by induction over the build up of formulas in a straightforward way

- Observe that in this interpretation the axiom **normal** will always be true

We get the completeness theorem for the Kripke logic — the derivable formulas are exactly those true in all Kripke models. The completeness is readily transferred to some of the other systems

T: Assume the schema $\Box F \to F$ is true in all maximal consistent sets. Then the binary relation $\{F | \Box F \in \mathcal{U}\} \subseteq \mathcal{V}$ is reflexive. The system T is complete for reflexive frames.

K4: Assume the schema $\Box F \to \Box\Box F$ is true in all maximal consistent sets. Then the binary relation is transitive and we have completeness for transitive frames.

S5: We get completeness for equivalence relations.

Gödel-Löb: Completeness for wellfounded and transitive frames

In all these cases we have a universal frame built up by the maximal consistent sets — and the extra axioms give geometrical conditions on the frame. For example

$\Box F \to F$**:** The frames are reflexive.

$\Box F \to \Box\Box F$**:** The frames are transitive.

1.5 Finiteness

The set of all maximal consistent set of formulas is huge. We can get finite versions of the above by considering maximal consistent subsets of all subformulas of a formula.

Gentzen systems

2.1 Strategies in and-or-trees

We have analyzed sequents Γ in propositional logic using sequent calculus by

- make a tree with sequents at the nodes

- at the root we have Γ

- at the leaf nodes there are only literals

- if all leaf nodes are axioms we have a derivation of Γ

- if one of the leaf nodes is not an axiom we get a falsification of Γ

Let us make the situation a little more abstract

- We have a finite tree with disjunctive nodes

- The leaf nodes are either colored blue or red

- We try to find a red colored leaf node

In modal logic we shall generalize this to

- We have a finite tree with disjunctive and conjunctive nodes

- The leaf nodes are either colored blue or red

7

- We try to find a strategy such that no matter which conjunctive branch is chosen we get to a red leaf node

2.2   The language of modal logic

- Propositional variables

- Literals

- Connectives: $\wedge$ $\vee$ $\square$ $\diamond$

Note that we do not have quantifiers. The negation is — as usual defined for literals — and extended to all formulas using double negation and de Morgan laws. The modal connectives are

Necessity: Unary connective $\square$ — called *box*

Possibility: Unary connective $\diamond$ — called *diamond*

Negation for modal connectives is given by

$$\neg\square F = \diamond\neg F$$
$$\neg\diamond F = \square\neg F$$

In our development of modal systems we have ingredients

Sequents: Finite set of formulas built up from literals by the connectives

Analyzing ∧ ∨: Usual rules for propositional logic

Critical sequents: Sequents with no ∧ or ∨ outermost — so they are of form Λ, ◇Δ, □Γ where Λ are literals and ◇Δ and □Γ is shorthand for ◇C, . . . , ◇D and □G, . . . , □H

Elementary critical sequent: A critical sequent Λ, ◇Δ, □Γ where Γ is either empty or consists of a single formula.

Elementary critical sequent subsuming a critical sequent: Given a critical sequent Λ, ◇Δ, □Γ. Then the elementary critical sequents subsuming it are all sequents Λ, ◇Δ, □G where G is from Γ.

From critical to elementary critical: So far in the analyzing tree we have disjunctive nodes. The critical sequents can be brought to elementary critical sequents by using a conjunctive node and the conjuncts go to each of the elementary critical sequents subsuming it.

2.3   The basic modal system K

The elementary critical sequents Λ, ◇Δ, □Γ are of three kinds

Blue: Λ contains both a literal and its negation

Red: Not blue and Γ is empty

Black: Neither red nor blue — and will be analyzed further

The black elementary critical sequents are analyzed using the following **normal** rule

$$\frac{\Delta, G}{\Lambda, \Diamond\Delta, \Box G}$$

Using this rule we have the following analyzing tree over $\Box F \wedge \Box(F \to G) \to \Box G$

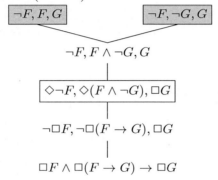

Here we have filled the nodes containing axioms with gray color. We have boxes around all critical elementary sequents

We have put the critical sequents in boxes and have colored the axioms blue. There is no way we can reach a white leaf — and the root sequence is derivable.

We can also derive the rule

$$\vdash F \Rightarrow \vdash \Box F$$

So we start with making an analyzing tree above $\Box F$. We have a critical sequent at the root and are lead to making an analyzing tree above F

The tree becomes a derivation if the tree above F is. Let us now go to an analyzing tree where we get a falsification. Consider the formula $\Box F \to \Box\Box F$. We get

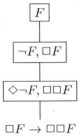

Here we have a strategy for getting to a white leaf. We can also read off a falsification. The three critical sequents give three worlds — and in each world the literals in the sequent are interpreted as false. We get

- U — world from $\Diamond\neg F, \Box\Box F$: no literals

- V — world from $\neg F, \Box F$: F true

- W — world from F : F false

The three worlds are related to each other by : $U \prec V$ and $V \prec W$

In the interpretations we have

World: Here we interpret the atomic formulas

Relation: Relations between the worlds

Frame: The graph given by the worlds and the relations between them

So far we have no branching from the critical sequents. This is so for $\Diamond F \to \Box F, \Diamond G$

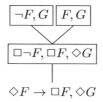

$$\Diamond F \to \Box F, \Diamond G$$

Here we also have three worlds

- U — world from $\Box \neg F, \Box F, \Diamond G$: no literals

- V — world from $\neg F$: F true and G false

- W — world from F : F false and G false

But the relation between the worlds are now : $U \prec V$ and $U \prec W$

2.4 Many worlds semantics

The short story about the semantics of the modal logic **K** is that it is the natural extension of propositional logic to analyzing tree with both conjunctive and disjunctive nodes. So let us see how this works out

Worlds: Falsification of a critical sequent $\Lambda, \Diamond\Delta, \Box\Gamma$. This is possible if the literals Λ does not both contain a literal and its negation.

Relating possible worlds: The critical sequents $\Lambda, \Diamond\Delta, \Box\Gamma$ with Γ not empty is analyzed by a conjunctive node — each of this may lead to new critical sequents and therefore new worlds

Propagating falsifications downwards: We propagate falsifications downwards from a premiss to the conclusion of the $\wedge$- and $\vee$-rules. In the rules for modal connectives we have a conjunctive node. If we have falsifications for all the premisses, we get a falsification for the conclusion.

Propagating falsifications upwards: From a falsification of the conclusion of an $\wedge$- or an $\vee$-rule, one of the premisses can be falsified. In the rules for modal connectives we get falsifications for all the premisses.

Given a sequent Θ we get

Worlds and relations: From the critical sequents and the conjunctive nodes

Falsification: Given by the red leaf nodes and the critical sequents involved in a strategy for reaching a red leaf starting from the root

A strategy gives a way to get from the root to a blue leaf node — for each disjunctive node one of the alternatives is chosen, for each conjunctive nodes all alternatives are considered.

The semantics for the modal connectives are

- $\Box F$ true in world W : F true in all worlds U related to W

- $\Diamond F$ true in world W : F true in some world U related to W

The leaf nodes have no nodes above them. So in a world from a leaf node all formulas $\Diamond D$ are false.

Now it is straightforward to transform the proof of completeness for propositional logic to a proof of the completeness of the modal logic **K**. The falsifications are finite worlds organized in a finite tree. It is an exercise to show that the modal logic **K** is complete for possible world semantics where the worlds are related as in a direct graph, but for the completeness we only need those worlds which can be organized as finite trees. We analyze falsifiability in **K** using trees with conjunctive and disjunctive nodes. We can then show that the complexity is PSPACE. In fact it is complete PSPACE.

2.5   Other systems

Some other rules for analyzing sequents are

Reflexivity $\dfrac{\Gamma, D, \Diamond D}{\Gamma, \Diamond D}$

Transitivity $\dfrac{\Diamond\Delta, \Delta, G}{\Lambda, \Diamond\Delta, \Box G}$

Well founded $\dfrac{\Diamond\Delta, \Delta, \neg\Box G, G}{\Lambda, \Diamond\Delta, \Box G}$

These rules correspond to extra assumptions about the relations between the worlds. Let us describe the modal systems.

T: We add the **reflexivity** rule to the system **K**. As before we get analyzing trees, but must now be careful how the rule is applied to ensure termination.

K4: Use the **transitivity** rule.

S4: Use the **reflexivity** and the **transitivity** rule.

GL: Use the **well founded** rule.

Note that the transitivity rule is a generalization of the normal rule. We get a larger sequent when we use the transitivity rule to analyze an elementary critical sequent. Adding the normal rule does not bring anything new — we do not get new axioms in the leaf nodes. In the same way the well founded rule is a generalization of both the transitivity rule and the normal rule.

We now had to look closer at the rules and the completeness theorems to find which frames these systems are complete for. We get:

K: Arbitrary frames — finite trees

T: Reflexive frames — finite reflexive trees

K4: Transitive frames — finite transitive trees

S4: Reflexive and transitive frames — finite reflexive, transitive frames

GL: Well founded frames — finite well founded trees

These are left as exercises. Note that in **S4** we do not get trees but only finite frames. This is exemplified by the McKinsey formula $\Box\Diamond F \to \Diamond\Box F$. We have the following analysis

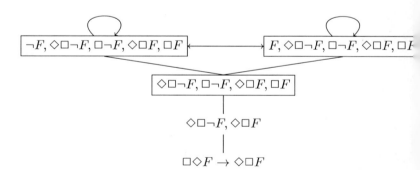

Instead of dublicating critical sequents in new nodes we have related them to previous critical sequents. But then we destroy the tree structure.

We do not need the lower critical sequent to get a falsification. We only need two related worlds — one where we falsify $\neg F$ and another where we falsify F.

We have also the modal system **S5**. Semantically it can be described as the system which is complete for frames which are symmetric, reflexive and transitive — we have en equivalence relation between the worlds. Syntactically it is easy to define a calculus for it. Here is a sketch

Sequents: Finite sets of formulas in modal logic. In our calculus this will correspond to worlds where we want to falsify the formulas in the sequent

Generalized sequents: Finite set of sequents separated by the sign |. The generalized sequents give all the worlds we want in the falsification.

Analysis of $\wedge$: From $\Gamma, F \wedge G | \Theta$ we branch off to $\Gamma, F | \Theta$ and $\Gamma, G | \Theta$

Analysis of $\vee$: From $\Gamma, F \vee G | \Theta$ to $\Gamma, F, G | \Theta$

Analysis of $\square$: From $\Gamma, \square F | \Theta$ to $\Gamma | \Theta | F$

Analysis of $\Diamond$: From $\Gamma, \Diamond F | \Theta$ to $\Gamma, \Diamond F, F | \Theta$

Propagation of $\Diamond$: From $\Gamma, \Diamond F | \Delta | \Theta$ to $\Gamma, \Diamond F | \Delta, \Diamond F | \Theta$

Axioms: A generalized sequent is an axiom if one of its sequents are

The proof of completeness is straightforward and is left as an exercise. The logically important point is that we analyze formulas in **S5** using only disjunctive nodes. The syntactic calculus is **NP** and not **PSPACE** as the other calculi are. There we need both conjunctive and disjunctive nodes.

Interpolation and diagonalization

3

3.1 Classical logic

For almost all rules in sequent calculus the symbols can be traced — from the conclusion to the premisses or from the premisses to the conclusion. The interpolation theorem uses this property. Let us start with a sequent

$$\Gamma, \Delta$$

We have divided it up into two parts — and to emphasize this division we write it as

$$\Gamma \circ \Delta$$

Definition 3.1 *Assume $\vdash \Gamma \circ \Delta$. A formula F is a separating formula of the sequent relative to the partition into Γ and Δ if*

- $\vdash \Gamma, F$

- $\vdash \Delta, \neg F$

- *F is built up from $\top$, $\bot$ and symbols occurring in both Γ and Δ — in particular the free variables of F are free in both Γ and Δ*

Γ is said to be the negative part of the partition and Δ is the positive part for the separating formula F.

Theorem 3.2 (Interpolation) *For any partition of a derivable sequent* $\vdash \Gamma \circ \Delta$, *we can find a separating formula for it.*

We are going to show that the separating formulas propagate through the proofs starting with the axioms and going through the cut free rules from the premisses to the conclusion. We shall look at individual rules. Observe that for all the rules — except cut — a partition of the conclusion induces in a natural way partitions of the premisses. Take as an example the rule

$$\frac{\Gamma, F \quad \Gamma, G}{\Gamma, F \wedge G}$$

Assume we have a partition of the conclusion into $\Delta \circ E, F \wedge G$. Then the two premisses are partitioned into $\Delta \circ E, F$ and $\Delta \circ E, G$. If we partition the conclusion into $\Delta, F \wedge G \circ E$, then the premisses are partitioned into $\Delta, F \circ E$ and $\Delta, G \circ E$.

Definition 3.3 *A rule is separable if for any partition of the conclusion and for any separating formulas of the induced partitions of the premisses, we can construct a separating formula of the conclusion. An axiom is separable if we can find a separating formula for any partition.*

We assume that there are no function symbols in our language. And then we show that the axioms and the cut free rules are separable and for the propagation of the separating formulas we have the following table

rule	negative part	positive part
$\wedge$	$\vee$	$\wedge$
$\vee$	$-$	$-$
$\forall$	$-$	$-$
$\exists$	$- \, / \, \forall$	$- \, / \, \exists$
$=$	$-$	$-$

There are many cases to consider, but they are all simple. We take some typical cases below. The remaining cases are similar to one of the cases we have written down.

Axiom which is split by the partition Assume that we have $\vdash \Gamma, L \, \circ \, \neg L, \Delta$. Then we can use $\neg L$ as separating formula.

Axiom which is not split by the partition Assume that we have $\vdash \Gamma, t = t \, \circ \, \Delta$. We can use $\bot$ as separating formula.

$\wedge$-*rule in the negative part* We have a partition $\vdash \Gamma, F \wedge G \circ \Delta$ and assume there are separating formulas I and J of the premisses with

$$\vdash \Gamma, F, I \qquad\qquad \vdash \Delta, \neg I$$

$$\vdash \Gamma, G, J \qquad\qquad \vdash \Delta, \neg J$$

Then by ordinary rules we get

$$\vdash \Gamma, F \wedge G, I \vee J \qquad\qquad \vdash \Delta, \neg(I \vee J)$$

and $I \vee J$ is a separating formula of the conclusion.

$\vee$-*rule or* $\forall$-*rule in the negative part* We can use the separating formula of the premiss as a separating formula of the conclusion. In the $\forall$-rule it is essential that the separating formula does not contain more free variables than are common to both parts.

$\exists$-*rule in the negative part* This is slightly complicated and we must use the assumption that we do not have function symbols. Then all terms are variables. Assume that we have a partition of the premisse $\vdash \Gamma, Ft \circ \Delta$ and there are separating formula I with

$$\vdash \Gamma, Ft, I \qquad\qquad \vdash \Delta, \neg I$$

Here we must be careful with the free variable t. If t does not occur in I we can use I as a separating formula of the conclusion. The same if t occurs both in $\Gamma, \exists xFx$ and Δ. The case when t occurs in I (as It) and in Δ but not in $\Gamma, \exists xFx$ remains. Then we use that t is a variable and get

$$\vdash \Gamma, \exists xFx, \forall yIy \qquad\qquad \vdash \Delta, \neg\forall yIy$$

$=$-*rule which is separated by the two parts* Assume we have partition $\vdash \Gamma, \neg s = t \circ Rt, \Delta$ with separating formula It. If t does occur in Rs, Δ, then we can use It as separating formula of the conclusion. Else we have

$$\vdash \Gamma, \neg s = t, It \qquad\qquad \vdash \Delta, Rt, \neg It$$

By $=$-rule $\vdash \Gamma, \neg s = t, Is$. Since t does not occur in Δ and is a free variable since there are no function symbols we can substitute s for t and get $\vdash \Delta, Rs, \neg Is$

$=$-*rule which is one of the parts* The separating formula of the premisse can also be used in the conclusion.

The remaining cases Similar to one of the cases above.

The case where we have function symbols in the language must be treated separately. There are various ways to do it, but we leave that for the reader to explore. We have included here a treatment of predicate logic. This is done for having a more complete treatment of interpolation, but we only need to consider the propositional logic within this book.

3.2 Modal logic

Consider now propositional modal logic. There we have two new operators $\Box$ and $\Diamond$ and new rules for them. To get interpolation theorem for modal logic we must look closer at the new rules.

Consider first the basic modal logic **K**. We must show that the following extra rule is separable

$$\frac{\Gamma, D}{\Lambda, \Diamond\Gamma, \Box D} \quad \text{and} \quad \frac{\Gamma}{\Lambda, \Diamond\Gamma}$$

There are two cases for partition of the conclusion.

$\Box D$ belongs to the negative part: So we have partition

$$\Lambda_1, \Diamond\Gamma_1, \Box D \circ \Lambda_2, \Diamond\Gamma_2$$

This partition gives the following partition of the premiss

$$\Gamma_1, D \circ \Gamma_2$$

We assume we have an interpolating formula I for the premiss

$$\vdash \Gamma_1, D, I \quad \text{and} \quad \vdash \Gamma_2, \neg I$$

and $\diamond I$ is an interpolating formula for the conclusion: (observe $\neg \diamond I = \square \neg I$)

$$\vdash \Lambda_1, \diamond \Gamma_1, \square D, \diamond I \quad \text{and} \quad \vdash \Lambda_2, \diamond \Gamma_2, \neg \diamond I$$

$\square D$ **belongs to the positive part**: Similar to the above. The interpolating formula is now $\square I$.

Theorem 3.4 *The interpolation theorem holds for the modal logics* ***K, K4, S4, GL***.

This follows as soon as we prove that the new rules are separable. Let us do it for **GL**. There we have the rule

$$\frac{\diamond \Gamma, \Gamma, \neg \square D, D}{\diamond \Gamma, \square D}$$

So assume that we have a partition of the conclusion in $\diamond \Gamma_1, \square D$ and $\diamond \Gamma_2$. We assume further that we have an interpolating formula I with

$$\vdash \diamond \Gamma_1, \Gamma_1, \neg \square D, D, I \qquad \vdash \diamond \Gamma_2, \Gamma_2, \neg I$$

But then we get

$$\vdash \diamond \Gamma_1, \diamond I, \square D \qquad \vdash \diamond \Gamma_2, \neg \diamond I$$

and $\Diamond I$ is an interpolating formula. If we partition the conclusion so that $\Box D$ comes in the other part, then $\Box I$ will be the interpolating formula.

3.3 Fixpoint theorem

We now want to prove the fixpoint theorem within **GL**. To simplify our exposition we introduce

$$\boxdot F = \Box F \wedge F$$

So $\boxdot F$ is true in a world if it is true there and in all later worlds. We say that G is *modalized* in $\mathcal{F}(G)$ if all occurrences of G in $\mathcal{F}(G)$ occurs within the scope of a modal operator.

Lemma 3.5 *Assume the atomic formula G is modalized in $\mathcal{F}(G)$ and let G' be a new atomic formula. Then*

$$\vdash \boxdot(\mathcal{F}(G) \leftrightarrow G) \wedge \boxdot(\mathcal{F}(G') \leftrightarrow G') \rightarrow (G \leftrightarrow G')$$

Proof. Suppose we have given a transitive, conversely well-founded frame where $\mathcal{F}(G) \leftrightarrow G$ and $\mathcal{F}(G') \leftrightarrow G'$ are true in all points. We want to prove by induction over the worlds that $G \leftrightarrow G'$ are always true starting with the topmost worlds. In a topmost world any formula $\Box H$ is true and any formula $\Diamond H$ is false. So since G and G' are modalized we must have the same truth value for $\mathcal{F}(G)$ and $\mathcal{F}(G')$ and hence for G and G'. Consider now a world w lower down in the frame and assume G and G' have the same truthvalue in the worlds above. But then

$\mathcal{F}(G)$ and $\mathcal{F}(G')$ have the same truthvalue in w since the only reference to the truthvalues of G and G' are from worlds above w. Hence G and G' have the same truthvalue in w. ■

Theorem 3.6 (Fixpoint) *Suppose the atom G is modalized in $\mathcal{F}(G)$. Then there is a formula H in the language of $\mathcal{F}$ and no G with $\vdash \Box(\mathcal{F}(G) \leftrightarrow G) \to (G \leftrightarrow H)$*

Proof. By the lemma we have

$$\vdash \Box(\mathcal{F}(G) \leftrightarrow G) \wedge \Box(\mathcal{F}(G') \leftrightarrow G') \to (G \to G')$$

By the interpolation theorem there is H in the language of $\mathcal{F}$ with no G such that

$$\vdash \Box(\mathcal{F}(G) \leftrightarrow G) \to (G \to H)$$
$$\vdash \Box(\mathcal{F}(G') \leftrightarrow G') \to (H \to G')$$

which gives the theorem. ■

3.4   Cantors diagonal argument

Georg Cantor proved in 1873 that the real numbers are uncountable. His proof then used topological properties of the real numbers. Later he came up with the well known diagonal argument. We consider the real numbers between 0 and 1 written in decimal notation and assume they are countable. We can then write them down as

$$a_i = 0.a_i^0 a_i^1 a_i^2 a_i^3 a_i^4 a_i^5 a_i^6 a_i^7 a_i^8 a_i^9 a_i^{10} a_i^{11} \ldots$$

where each a_j^i is a decimal. We can write them down as an infinite matrix

a_0^0	a_0^1	a_0^2	a_0^3	a_0^4	a_0^5	a_0^6	a_0^7	a_0^8	a_0^9
a_1^0	a_1^1	a_1^2	a_1^3	a_1^4	a_1^5	a_1^6	a_1^7	a_1^8	a_1^9
a_2^0	a_2^1	a_2^2	a_2^3	a_2^4	a_2^5	a_2^6	a_2^7	a_2^8	a_2^9
a_3^0	a_3^1	a_3^2	a_3^3	a_3^4	a_3^5	a_3^6	a_3^7	a_3^8	a_3^9
a_4^0	a_4^1	a_4^2	a_4^3	a_4^4	a_4^5	a_4^6	a_4^7	a_4^8	a_4^9
a_5^0	a_5^1	a_5^2	a_5^3	a_5^4	a_5^5	a_5^6	a_5^7	a_5^8	a_5^9
a_6^0	a_6^1	a_6^2	a_6^3	a_6^4	a_6^5	a_6^6	a_6^7	a_6^8	a_6^9
a_7^0	a_7^1	a_7^2	a_7^3	a_7^4	a_7^5	a_7^6	a_7^7	a_7^8	a_7^9
a_8^0	a_8^1	a_8^2	a_8^3	a_8^4	a_8^5	a_8^6	a_8^7	a_8^8	a_8^9
a_9^0	a_9^1	a_9^2	a_9^3	a_9^4	a_9^5	a_9^6	a_9^7	a_9^8	a_9^9

Here we have indicated the diagonal as a line. From the diagonal we construct a new number

$$d = 0.d_0 d_1 d_2 d_3 d_4 \ldots$$

where each decimal of d is different from the diagonal

$$d_i = \begin{cases} 3 & \text{,if } a_i^i \neq 3 \\ 7 & \text{,if } a_i^i = 3 \end{cases}$$

If the matrix contains all real numbers between 0 and 1, then the d must be one of them and looking where the row with d intersects the diagonal we get a contradiction.

The ingredients in Cantors argument are

- Construct the matrix

- Transform the diagonal so that it looks like a row

- Consider the place where the row intersects the diagonal — this is called the critical point

There are two main outcomes of the argument

- There is no row corresponding to the transformed diagonal. There are too few rows, and this is shown by looking at the critical point.

- There is a row corresponding to the transformed diagonal. The critical point will be a fixed point.

3.5 Russells paradox

We construct the matrix with sets along the axis and where row x intersects column y we put the truth value of $x \in y$. As transformation we use negation $\neg$. The diagonal defines the Russell set $\mathcal{R}$ given by the truth values of $\neg x \in x$. We get a contradiction if we assume there is a row corresponding to $\mathcal{R}$. The problem comes at the critical point. There we would have

$$\mathcal{R} \in \mathcal{R} \Leftrightarrow \mathcal{R} \notin \mathcal{R}$$

which is impossible.

3.6 Fixed point of functions

We consider functions with one argument. They are the point on the axis. The matrix is constructed by composing the functions. The diagonal is

$$\lambda f.\lambda x.f(f(x))$$

Assume now that we have a transformation Φ of such functions with one argument. The transformed diagonal is

$$\lambda f.\Phi(ff)$$

Assume now that the transformed diagonal correspond to the function g. Then at the critical point we get

$$\Phi(gg) = gg$$

and gg is the fixed point of Φ. We can write it as

$$(\lambda f.\Phi(ff))(\lambda f.\Phi(ff))$$

3.7 Coding

We use codes in a datastructure to represent syntax and logical calculi over the datastructure to do syntactical transformation. We have found the following to be most useful

- As datastructure we use binary trees $\mathcal{B}$. They are given by

  - Constant — **nil** : $\mathcal{B}$ — the empty tree
  - Binary constructor — **cons** : $\mathcal{B} \times \mathcal{B} \to \mathcal{B}$
  - Binary predicate — $\prec$: $\mathcal{B} \times \mathcal{B} \to$ **Boole** — x constructed before y

- Syntax — can be expressed with a Δ_0-formula

- Basic calculus — sufficient to prove all true Δ_0-sentences

- Extended calculus — includes also induction over Δ_0-formulas

Often the datastructure is the unary numbers. But then we had to include more functions to make the coding work.

It is important that we distinguish between the code of an object and the object itself. The following notations will be used

- $\{p\}(x)$ — p is a programcode for a function, x is the input and $\{p\}(x)$ is the result of applying the program p to x

- $\lceil A \rceil$ — the code for the formula A

With the programcode we have some new notations

- $\{p\}(x) \uparrow$ — $\{p\}(x)$ does not terminate

- $\{p\}(x) \downarrow$ — $\{p\}(x)$ terminates

- $\{p\}$ total — it terminates for all inputs

- $\{p\}$ partial — may or may not terminate

- $\{p\} \cong \{q\}$ — terminate for exactly the same inputs and there have the same value

3.8 The halting problem

We build a matrix of programcodes and inputs — both given as binary trees. On row x and column y we put $\{x\}(y)$. If we could decide whether $\{x\}(y)$ terminates, then transform the diagonal by switching termination and non-termination. There is a row corresponding to the transformed diagonal and at the critical point we get a contradiction. We conclude that the halting problem is not decidable.

3.9   The fix point theorem

This is the coded version of the argument above of fix point of functions. So assume we have a total function h. The matrix is given by partial functions $\{x\}$ and $\{y\}$ where at the intersection we have the composition of them. This can be written as $\{x \circ y\}$. Now to the details of the diagonal construction

- the diagonal is d — it takes as argument a code and produces a code and is itself a code

- the transformed diagonal $h(d)$ corresponds to row $\{g\}$

- the critical point gives the fix point — $\{h(g \circ g)\} \cong \{g \circ g\}$

- the fix point — $\{g \circ g\} \cong \{h(d) \circ h(d)\} \cong \{(\lambda x.h(x \circ x)) \circ (\lambda x.h(x \circ x))\}$

3.10   Diagonal lemma

We now consider formulas with free variable x over the language of our datastructure. A formula Ax and a formula (or sentence) B gives the sentence $A\lceil B\rceil$ — substitute the code of B for the free variable x in Ax. This form of composition can be used in a diagonal argument.

Given a formula Fx with free variable x. There is then a sentence B with $B \leftrightarrow F\lceil B\rceil$.

Matrix: We are in the language of binary trees. To row y and column z we get the binary tree $\mathbf{sub}(y, z)$ — the result of considering y as a code for a formula Yx with free variable x and then substitute z for x.

Diagonal: $\lambda y.\mathbf{sub}(y, y)$

Transformed diagonal: $A(\mathbf{sub}(x, x))$ — corresponds to a row

Critical point: Gives the sentence B

All the coding / uncoding needed here are purely syntactical and can be done as Δ_0-formulas.

Provability

4

4.1 Expressing syntax

In provability we try to express truth by using syntactical constructions. We start with axioms as something obviously true and use syntactical rules to get from old truths to new truths. This is an old idea — dating back to Aristotle and to Euclid. To make it precise we must decide how to represent syntax. We propose the following

Language: We have the language of binary trees $\mathcal{B}$ with

- **nil** : $\mathcal{B}$
- **cons** : $\mathcal{B} \times \mathcal{B} \rightarrow \mathcal{B}$
- $\prec$: $\mathcal{B} \times \mathcal{B} \rightarrow$ **Boole**

Syntactical construction: Something that can be described by a Δ_0-formula in $\mathcal{B}$. This is a formula which only contains bounded quantifiers.

Calculus: Above the datastructure $\mathcal{B}$ there are three proposals for calculi

    Basic calculus: We add to predicate logic with equality all true Δ_0-sentences

    Skolem calculus: In addition — induction over Σ_1-formulas

    Peano-calculus: In addition — induction over all formulas

The pioneer here was Thoralf Skolem with his 1923-paper "Begründung der elementaren Arithmetik durch die rekurrierende Denkweise ohne Anwendung scheinbarer Veränderlichen mit unendlichem Ausdehnungsbereich." He had read Russell and Whiteheads Principia Mathematicae and was irritated by their founding of simple arithmetical truths. Instead of founding everything on (higher order) logic he introduced — in modern terminology — the following

- The datastructure of unary numbers

- Primitive recursive definitions as a programming language over the datastructure

- A programming logic over the datastructure showing correctness of simple truths

If we add primitive recursive functions to the language it is sufficient to consider induction over quantifier free formulas — and induction over Σ_1-formulas is derivable. With not enough functions in the language it is better to start with induction over Σ_1-formulas. In any case it seems to be the thing that is needed in this chapter.

There is a criticism against the datastructure of unary numbers. It is a poor datastructure to express syntax. In fact we have

- natural numbers with $+$ is decidable (Presburger, Skolem)

- natural numbers with $\times$ is decidable (Skolem)

We must use natural numbers with both $+$ and $\times$ to get Gödels incompleteness, and then the coding needs the extra trick of Gödels β-function and the Chinese remainder theorem. Using the datastructure of binary trees avoids all these extra problems — and we know already from Lisp and other programming languages that this datastructure works well to simulate syntactical constructions.

4.2 Calculi of syntax

The datastructure of binary trees $\mathcal{B}$ gives a reasonable language for syntactical constructions. We have the following simple definitions:

$$
\begin{array}{rcl}
x \preceq y & : & x \prec y \vee x = y \\
\forall x \prec y.Fx & : & \forall x(x \prec y \rightarrow Fx) \\
\exists x \prec y.Fx & : & \exists x(x \prec y \wedge Fx) \\
\forall x \preceq y.Fx & : & Fy \wedge \forall x \prec y.Fx \\
\exists x \preceq y.Fx & : & Fy \vee \exists x \prec y.Fx \\
x = \mathbf{hd}(y) & : & \exists z \prec y.y = \mathbf{cons}(x,z) \\
x = \mathbf{tl}(y) & : & \exists z \prec y.y = \mathbf{cons}(z,x) \\
x \prec \mathbf{hd}(y) & : & \exists u \prec y.\exists v \prec y.(y = \mathbf{cons}(u,v) \wedge x \preceq v) \\
x \prec \mathbf{tl}(y) & : & \exists u \prec y.\exists v \prec y.(y = \mathbf{cons}(u,v) \wedge x \preceq u) \\
x \preceq \mathbf{hd}(y) & : & x = \mathbf{hd}(y) \vee x \prec \mathbf{hd}(y) \\
x \preceq \mathbf{tl}(y) & : & x = \mathbf{tl}(y) \vee x \prec \mathbf{tl}(y) \\
x = \mathbf{hd^*}(y) & : & x \preceq y \wedge \forall z \preceq y(x \prec z \rightarrow x \preceq \mathbf{hd}(z)) \\
x = \mathbf{tl^*}(y) & : & x \preceq y \wedge \forall z \preceq y(x \prec z \rightarrow x \preceq \mathbf{tl}(z))
\end{array}
$$

Observe how far we come with constructions using bounded quantifiers. In particular we are able to define **hd*** and **tl*** in this way — and using this we get finite sequences of information, finite proofs, finite runs and so on using only bounded quantifiers. This is not possible in the datastructure of unary numbers without extra functions.

For the calculi we start with predicate calculus with equality and then add axioms for the datastructure $\mathcal{B}$. There are three main levels

Basic calculus: Add all true Δ_0-sentences

Skolem calculus: In addition — induction over all Σ_1-formulas

Peano calculus: In addition — induction over all formulas

Why these levels?

- A Δ_0-sentence is built up from literals using connectives and bounded quantifiers. The truth-value is calculated using a finite AND-OR tree above it.

- Induction is supposed to reflect the build up of the datastructure. Using induction involving unbounded $\forall$-quantifiers we have already assumed some knowledge of the totality of the datastructure. This seems to be a deeper assumption than the one for Σ_1-formulas.

As something intermediate between the basic calculus and the Skolem calculus we often use the Robinson axioms:

R1: $\neg\text{nil} = (x \cdot y)$

R2: $(x \cdot y) = (u, v) \to x = u \wedge y = v$

R3: $x = \text{nil} \vee \exists u, v . x = (u \cdot v)$

R4: $\neg x < \text{nil}$

R5: $x < (u \cdot v) \leftrightarrow x = u \vee x < u \vee x = v \vee x < v$

It is an exercise to show that they are intermediate — we can derive all true Δ_0-sentences for them, and they can all be derived in Skolem calculus. This is left to the reader.

4.3 Gödel - Löb modal logic

We are in a system strong enough to express syntax and proving syntactical transformation. We have a coding of syntax — formulas and proofs. Let F be a formula. Then $\Box F$ means that there is a code of a proof of F. We shall look further into the requirements of the code by looking at the properties of $\Box F$. In fact we shall prove

$$
\begin{array}{ll}
\textbf{GL 1} & \vdash F \Rightarrow \vdash \Box F \\
\textbf{GL 2} & \vdash \Box F \wedge \Box(F \to G) \to \Box G \\
\textbf{GL 3} & \vdash \Box F \to \Box\Box F \\
\textbf{GL 4} & \vdash \Box(\Box F \to F) \to \Box F
\end{array}
$$

Necessitation — GL 1

Assume $\vdash F$. Then the coding is such that we can translate this step for step into a proof of $\vdash \Box F$. And we have

$$\vdash F \;\Rightarrow\; \vdash \Box F$$

Normality — GL 2

If we have proofs of $F \to G$ and F, then we can transform this into a proof of G by using modus ponens. This transformation is done without analyzing the details of the two proofs. This gives the derivation of

$$\vdash \Box(F \to G) \to (\Box F \to \Box G)$$

There are other notions of proofs where **GL 2** is far from obvious

- it is a direct proof of F

- it is the shortest possible proof of F

Fair coding

So far the coding is just new names for the syntactical elements. We want the coding to be such that

Fair coding of equality: $\forall x, y.(x = y \to \Box x = y)$

Fair coding of inequality: $\forall x, y.(x \neq y \to \Box x \neq y)$

These are reasonable extra conditions on the coding.

Σ_1-completeness

Theorem 4.1 (Σ_1 completeness) *Assume that the theory contains Σ_1-induction and we have a fair coding of equality and inequality. Then for all Σ_1 formulas G*

$$\vdash G \to \Box G$$

Here G may contain free variables.

Proof. The proof is first by induction over the build up of Δ_0 formulas. Then we show that the principle still holds if we have $\exists$-quantifiers outermost.

Literals: The fair coding of equality and of inequality gives the principle for $x = y$ and for $x \neq y$. It also gives the principle for $x = (u \cdot z)$ and $x \neq (u \cdot z)$. Just substitute $(u \cdot z)$ for y. For $x < y$ we use Σ_1 induction over y. In the induction start $y = \mathbf{nil}$ we have $\vdash \neg x < \mathbf{nil}$ and trivially the principle. Assume the principle true for y and z. We then get it for $(y \cdot z)$ by using:

$$x < (y \cdot z) \leftrightarrow x = y \lor x < y \lor x = z \lor x < z$$

Now to $\neg x < y$. Again we use Σ_1-induction over y. For the induction start $y = \mathbf{nil}$ we note

$$\vdash \Box \neg x < \mathbf{nil}$$

In the induction step we again use the equivalence above.

Conjunction: Assume $\vdash F \to \Box F$ and $\vdash G \to \Box G$. But from **GL0** we have $\vdash \Box(F \to (G \to F \wedge G))$ and using **GL1** and propositional logic we get $\vdash F \wedge G \to \Box F \wedge G$.

Disjunction: Here we use **GL0** with $\vdash \Box(F \to F \vee G)$ and $\vdash \Box(G \to F \vee G)$.

Bounded quantifiers: We use Σ_1-induction over y to prove

$$\vdash \exists x < y.Gx \to \Box \exists x < y.Gx$$

Note that the formula $\exists x < y.Gx$ is Δ_0 and hence that the whole formula is Σ_1.

In the same way we prove by induction

$$\vdash \forall x < y.Gx \to \Box \forall x < y.Gx$$

We conclude that the principle is true for all Δ_0 formulas. Now we note that it can be extended with $\exists$-quantifiers in front.

Existential quantifier: We assume $\vdash Fx \to \Box Fx$ for arbitrary x. Furthermore by **GL0** $\vdash \Box(Fx \to \exists y.Fy)$. Then $\vdash Fx \to \Box \exists y.Fy$ and $\vdash \exists y.Fy \to \exists y.Fy$. ∎

So the principle is true for all Σ_1 formulas. In particular it is true for the Σ_1 formula $\Box F$ and we get for theories with Σ_1-induction

GL2: $\vdash \Box F \to \Box \Box F$

Löbs rule — $\vdash \Box S \to S \Rightarrow \vdash S$

Here we use **GL0**, **GL1** and **GL2** and in addition the fix point theorem. From the fix point theorem there is an I with $\vdash I \leftrightarrow (\Box I \to S)$. We then have

$$\vdash I \to (\Box I \to S)$$
$$\vdash \Box I \to (\Box\Box I \to \Box S)$$
$$\vdash \Box I \to \Box S$$
$$\vdash \Box S \to S \text{ , assumption}$$
$$\vdash \Box I \to S$$
$$\vdash I$$
$$\vdash \Box I$$
$$\vdash S \text{ , conclusion}$$

Löbs axiom — GL3

We abbreviate $B = \Box(\Box F \to F)$, $C = \Box F$ and $D = B \to C$. Then

$$\vdash \Box D \to (\Box B \to \Box C)$$
$$\vdash B \to (\Box C \to C)$$
$$\vdash B \to \Box B \text{ , since } B \text{ starts with } \Box$$
$$\vdash \Box D \to (B \to C)$$
$$\vdash \Box D \to D$$
$$\vdash D \text{ , by Löbs rule}$$

And we are done. Note that we had to use Σ_1-induction to prove **GL2** and this was again used in the proof of **GL3**.

Deriving GL2 from the rest

$\vdash \Box(\Box(\Box F \land F) \to (\Box F \land F)) \to \Box(\Box F \land F)$ by **GL3**
$\vdash \Box(\Box F \land F) \to \Box\Box F$ by **GL0** and **GL1**
$\vdash \Box(\Box F \land F) \to \Box F$ by **GL0** and **GL1**
$\vdash \Box(\Box(\Box F \land F) \to (\Box F \land F)) \to \Box\Box F$ by logic
$\vdash F \to (\Box(\Box F \land F) \to (\Box F \land F))$ by logic
$\vdash \Box F \to \Box(\Box(\Box F \land F) \to (\Box F \land F))$ by **GL0** and **GL1**
$\vdash \Box F \to \Box\Box F$ by logic

Gödel interpreted provability as a modal operator — $\Box F$ means that F is provable. This makes perfectly good sense — also for complex modal formulas — if our logic system contains enough to handle syntactical construction. We shall prove that with this interpretation the valid formulas of the Gödel-Löb logic are true.

5

Incompleteness

5.1 Gödels zeroth incompleteness theorem

The theory of binary trees cannot treat the true sentences in a reasonable way. We have

- The language of binary trees is rich enough to express syntax with Δ_0-formulas and rich enough to express provability and computability with Σ_1-sentences.

- The simplest calculus on it makes all true Σ_1-sentences provable.

- There is no calculus on it making all true Π_1-sentences provable.

If there were a calculus for all true Π_1-sentences, then we would get a decision procedure for the halting problem.

5.2 Gödels first incompleteness theorem

In the theory of binary trees we we have under quite general assumptions

- all true Δ_0-sentences are provable

- only true sentences are provable

- provability is only partially computable

45

we conclude that there must be a true Π_1-sentence which is not provable. Here we shall show a way to get around the second assumption, and let the incompleteness be a more syntactical matter. Gödel considered two notions

- A theory is *consistent* if $\nvdash \bot$.

- A theory is *ω-consistent* if there are no Fx with $\vdash \exists x.Fx$ and $\vdash \neg Fp$ for all p

If a theory is ω-consistent, then there are no F with $\vdash F$ and $\vdash \neg F$. Instead of ω-consistency it may be more perspicuous to use the following consequence

Lemma 5.1 *If the theory is ω-consistent and provability is Σ_1, then it is 1-consistent — that is*

$$\vdash \Box F \Rightarrow \vdash F$$

Proof. For assume $\vdash \Box F$ or $\vdash \exists p.\textbf{PROOF}(p, \lceil F \rceil)$. By ω-consistency there must be q with $\vdash \textbf{PROOF}(q, \lceil F \rceil)$. But this is a Δ_0 sentence and hence true. Therefore $\vdash F$. ∎

Theorem 5.2 (First incompleteness) *Assume we have a theory in the language of pairs where all true Δ_0 sentences are provable. Let G be a sentence such that*

$$\vdash G \leftrightarrow \neg \Box G$$

If the theory is consistent, then $\nvdash G$. If it is in addition 1-consistent, then $\nvdash \neg G$

Proof.

Assume that the theory is consistent. Assume $\vdash G$. Then $\vdash \Box G$ by **GL0** and from the definition of G we get $\vdash \neg G$ contradicting consistency.

Assume that the theory is 1-consistent. Assume $\vdash \neg G$. Then $\vdash \Box G$ and from the above by 1-consistency $\vdash G$. This contradicts the consistency of the theory. ■

Under the weak assumptions above we get a sentence G such that neither it nor its negation $\neg G$ is provable. So the theory is not complete.

The assumption used before about the theory being partially computable is here reflected in the assumption that $\Box F$ is a Σ_1 formula.

5.3 Gödels second incompleteness theorem

We now use the stronger assumption **GL2** of provability.

Theorem 5.3 (Second incompleteness) *Let G be the sentence used in the first incompleteness theorem. Then*

$$\vdash G \leftrightarrow \neg\Box\bot$$

Proof. We have $\vdash G \rightarrow \neg\Box G$. Then use $\vdash \bot \rightarrow G$ and **GL0** and **GL1** to get $\vdash \Box\bot \rightarrow \Box G$. Hence $\vdash G \rightarrow \neg\Box\bot$ which is half of the equivalence.

Conversely by **GL2** we have $\vdash \Box G \rightarrow \Box\Box G$ and hence $\vdash \Box G \rightarrow \Box\neg G$. Then $\vdash \Box G \rightarrow \Box(G \wedge \neg G)$ and $\vdash \Box G \rightarrow \Box\bot$ and $\vdash \neg G \rightarrow \Box\bot$ which is the other half of the equivalence. ■

Both incompleteness theorems use that provability is $\mathbf{\Sigma}_1$. The first incompleteness theorem requires that all true $\mathbf{\Delta}_0$ sentences are provable, while the second incompleteness theorem requires $\mathbf{\Sigma}_1$ induction in the theory — to get **GL2** — and that the coding represents equality and inequality in a fair way.

The second incompleteness theorem shows that it makes sense to talk about the Gödel sentence of a theory — and that this sentence is equivalent to $\neg\Box\bot$. The actual theories enters only through the representation of $\Box$.

5.4 Tarskis theorem

Assume that we have a formalized notion of truth one within our theory. That is there is a predicate **TR** such that for all sentences S

$$\vdash \mathbf{TR}(\lceil S \rceil) \leftrightarrow S$$

The point is that the S on the left hand side occurs as the coded representation of it. We can then use the fix point theorem to get a sentence T with

$$\vdash \neg\mathbf{TR}(\lceil T \rceil) \leftrightarrow T$$

and we have an immediate contradiction.

Theorem 5.4 (Tarski) *There is no formalized theory of truth in a theory where we can prove the fix point theorem.*

Provability logic

6

6.1 Solovays first completeness theorem

In this section we shall show — following work of Robert Solovay — that GL tells the whole story about provability in elementary theories with Σ_1-induction.

We want to simulate trees and interpretations over a tree in an elementary theory. So assume that we have a representation of the tree given by a downmost node 0 and the accessibility relation $\prec$ and downmost world 0. For notational simplicity we use natural numbers and usual less than relation on them. Define a primitive recursive (or using Σ_1 induction) climbing function h by

$$
\begin{aligned}
h(0) &= 0 \\
h(x+1) &= \left\{
\begin{array}{ll}
j & \text{, where } j \succ h(x) \\
& \text{and } x \text{ proves } \exists y > x \cdot h(y) \succ j \\
h(x) & \text{, otherwise}
\end{array}
\right.
\end{aligned}
$$

So it describes a man climbing up the frame, and he can only climb up a step if he can prove that he will not stay there. (Or, it could be a refugee who is allowed to enter a country only if he can prove that he will go to a more favorable country.) We then define predicates S_i expressing that the climber will ultimately end up with world i

$$
S_i = \exists x \forall y > x \cdot h(y) = i
$$

We can then prove

$$\vdash S_i \rightarrow \Box \neg S_i \qquad \text{for } i \succ 0$$
$$\vdash S_i \rightarrow \neg \Box \neg S_j \qquad \text{for } j \succ i$$
$$\vdash \neg(S_i \wedge S_j) \qquad \text{for } i \neq j$$
$$\vdash \bigvee_i \cdot S_i$$
$$\vdash S_i \rightarrow \Box \bigvee_{j \succ i} S_j \qquad \text{for } i \succ 0$$

Only the last requires some work. We can derive within the theory using $i \succ 0$ and the formalized $\mathbf{\Sigma_1}$-completeness

$$S_i \rightarrow \exists a \cdot h(a) = i$$
$$\exists a \cdot h(a) = i \rightarrow \bigvee_{j \succeq i} S_j$$
$$\Box \exists a \cdot h(a) = i \rightarrow \Box(S_i \vee \bigvee_{j \succ i} S_j)$$
$$\exists a \cdot h(a) = i \rightarrow \Box \exists a \cdot h(a) = i \quad \text{using } \mathbf{\Sigma_1}\text{-completeness}$$
$$S_i \rightarrow \Box(S_i \vee \bigvee_{j \succ i} S_j) S_i \rightarrow \Box(\neg S_i)$$
$$S_i \rightarrow \Box \bigvee_{j \succ i} S_j$$

We are now ready to simulate any finite, transitive frame in arithmetic. So assume such a frame is given and let S_i be the corresponding sentences. There is also defined a provability operator. To any formula F in GL we define a formula $F^\star$ in arithmetic by

- if P is an atomic formula, then $P^\star = \bigvee\{S_i | P \text{ is true in } i\}$

- $(P \wedge Q)^\star = P^\star \wedge Q^\star \quad (P \vee Q)^\star = P^\star \vee Q^\star$

- $(\Box P)^\star = \Box P^\star$

We say that $P^\star$ interprets P. This is justified by

Lemma 6.1
$$i \models P \;\Rightarrow\; \vdash S_i \to P^\star$$
$$i \not\models P \;\Rightarrow\; \vdash S_i \to \neg P^\star$$

Proof. We prove this by induction over the formula P. Observe that it is true for atomic formula and it is preserved by Boolean combinations. We are left to prove it for formula $\Box Q$ assuming it true for Q. So assume first $i \models \Box Q$. Then

$$\forall j \succ i \cdot j \models Q$$
$$\forall j \succ i \cdot \vdash S_j \to Q^\star$$
$$\vdash \bigvee_{j \succ i} S_j \to Q^\star$$
$$\vdash \Box \bigvee_{j \succ i} S_j \to \Box Q^\star$$
$$\vdash S_i \to \Box Q^\star$$

And assume $i \not\models \Box Q$. Then

$$\exists j \succ i \cdot j \models Q$$
$$\exists j \succ i \cdot \vdash S_j \to \neg Q^\star$$
$$\exists j \succ i \cdot \vdash \neg \Box \neg S_j \to \neg \Box Q^\star$$
$$\vdash S_i \to \neg \Box Q^\star$$

∎

Theorem 6.2 (Solovays first completeness theorem) $\vdash_{GL} A \Leftrightarrow \forall \star \vdash_S A^\star$

Proof. We have already proved the implication $\Rightarrow$. To the other way assume $\not\vdash_{GL} A$. There exists then a finite, transitive, conversely wellfounded frame with downmost element 1 and $1 \not\models A$. Tack on a new element 0 in the frame below 1 and let the interpretations in 0 be arbitrary. We have

$$\vdash S_1 \rightarrow A^\star$$
$$\vdash \neg\square\neg S_1 \rightarrow \neg\square A^\star$$
$$\vdash S_0 \rightarrow \neg\square\neg S_1$$
$$\vdash S_0 \rightarrow \neg\square A^\star$$

But S_0 is true (even if it is not provable). Therefore $\square A^\star$ is false and hence $\not\vdash_S A^\star$. ∎

We have used Σ_1-induction. The same arguments goes through for any stronger system.

6.2   Solovays second completeness teorem

In the proof of Solovays first completeness theorem we wandered between provability and truth. This is going to be done even more so in the Solovays second completeness theorem. First we introduce a new logical system GLS — Gödel Löb Solovay

- all valid GL-sentences

- $\square A \rightarrow A$

- modus ponens

Note that we do not have the necessitation rule ($\vdash A \Rightarrow\vdash \square A$) in GLS. Let F be any formula. We are going to use the formula

$$\bigwedge\{\square A \to A | \square A \text{ is a subformula of } F\} \to F$$

This is written shortly as $\bigwedge(\square A \to A) \to F$.

Theorem 6.3 (Solovays second completeness theorem) *The following are equivalent*

1. $GL \vdash \bigwedge(\square A \to A) \to F$

2. $GLS \vdash F$

3. $\forall \star F^\star$ true

Proof. Here it is straightforward to prove $1 \Rightarrow 2$ and $2 \Rightarrow 3$. So assume $GL \nvdash \bigwedge(\square A \to A) \to F$. Then there is finite, transitive, conversely wellfounded frame with downmost element 1 giving a countermodel for it. Tack on a new node 0 below 1 and now we assume that 0 has exactly the same literals true as 1. Then $1 \nvDash F$ and $1 \vDash \bigwedge(\square A \to A)$. Let the sentences S_i be defined by the climbing function as in the proof of the first completeness theorem. We first show for subformulas of F

$$1 \vDash B \Rightarrow \vdash S_0 \to B^\star$$
$$1 \nvDash B \Rightarrow \vdash S_0 \to \neg B^\star$$

We use induction over B. If B is atomic, then $B^\star$ is a disjunction of S_is. Assume $1 \vDash B$. Then by construction of 0 we have $0 \vDash B$ and S_0 is a disjunct of $B^\star$. So $\vdash S_0 \to B^\star$. Now assume $1 \nvDash B$. By construction of 0 we have $0 \nvDash B$ and S_0 is not a disjunct of $B^\star$. So $\vdash S_0 \to \neg B^\star$. The properties are easily extended through Boolean combinations. It remains to prove it for $\Box C$ given the properties for C.

Assume $1 \models \Box C$. Then $\forall j \succ 1 \cdot j \models C$ and by properties proved in the first completeness theorem we have $\forall j \succ 1 \cdot \vdash S_j \to C^\star$. We now use that $1 \models \Box C \to C$ to conclude $\vdash S_1 \to C$. This gives $\vdash S_1 \vee \bigvee_{j \succ 1} S_j \to C^\star$. We now use $\vdash S_0 \vee S_1 \vee \bigvee_{j \succ 1} S_j$ to conclude $\vdash C^\star$ and $\vdash \Box C^\star$ and $\vdash S_0 \to \Box C^\star$.

Assume $1 \not\models \Box C$. Then $\exists j \succ 1 \vdash S_j \to \neg C^\star$ and $\exists j \succ 1 \vdash \neg \Box \neg S_j \to \neg \Box C^\star$. But $\vdash S_0 \to \neg \Box \neg S_j$. This gives $\vdash S_0 \to \neg \Box C^\star$.

This proves the properties. Now to the conclusion of the proof of the completeness theorem. We have assumed that $1 \not\models F$. But then $\vdash S_0 \to \neg A^\star$. But S_0 is true. Hence $A^\star$ is false. ∎

Let us give some simple applications. Assume we want to find a sentence S which is true but not provable in arithmetic. We then try to falsify $\neg S, \Box S$ in GL and get the following falsification tree

So we have two worlds — an upper with S false and a lower with S true. Observe now that in the lower one we have $\Box A \to A$ true for any subformula $\Box A$ of the sequent. There is only one $\Box F$ and we have $\Box F \to F$ true in the lower world. Now tack on a new world 0 below and we then have an interpretation for S.

We can interpret it as $\neg S_2$. So in arithmetic we have $\neg S_2$ true, but not provable.

What is the complexity of the interpreted sentences? We observe that for a topmost node j we can write $S_j = \exists x \cdot h(x) = j$ since we cannot climb further, and we get S_j a $\mathbf{\Sigma}_1$-sentence. For a node i lower down we can write

$$S_i = \exists x \cdot h(x) = i \wedge \bigwedge_{k \succ i} \neg \exists y \cdot h(y) = k$$

and we get S_i a Boolean combination of $\mathbf{\Sigma}_1$-sentences.

In our example we have a true $\mathbf{\Pi}_1$-sentence which is not provable.

Let us now try to find a sentence S such that neither S nor $\neg S$ is provable. We are then led to the following falsification tree

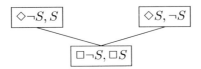

As before observe that in the lower world we have $\square A \to A$ true for all subformulas $\square A$ of the sequent. Tack on a new node below and as before we have an interpretation for S — and a formula which is neither provable nor its negation is provable. This is a variant of the Rosser construction.

It is surprising that the incompleteness phenomena can be expressed with the decidable modal logic GL. On the other hand it gives a limitation on our analysis so far. The second incompleteness theorem tells us that we cannot prove $\neg\Box\bot$. The system enters in our understanding of $\Box$, but this understanding is not expressed in the language of GL.

Multi modal logic

7.1 Many modal operators

A modal operator $\Box$ is *normal* if it satisfies the rules for system **K** — that is necessitation rule and the normal axiom. We may have many normal operators at the same time. The theory — in the Gentzen or the Frege version goes through as before. We write the modal operators and their transition relations as

$$[i] \; \langle i \rangle \; [j] \; \langle j \rangle \; \dots \xrightarrow{i} \xrightarrow{j}$$

The transitions $\xrightarrow{i}, \xrightarrow{j}, \xrightarrow{k}, \xrightarrow{l}$ have the (i, j, k, l)-confluence property if the following diagram is commutative (that is to every A, B, C there is a D such that $\dots$):

Theorem 7.1 *To the geometric property (i, j, k, l)-confluence corresponds the axiom*

$$\langle i \rangle [j] F \rightarrow [k] \langle l \rangle F$$

Proof.

$\Rightarrow$: Suppose that we have (i, j, k, l)-confluence. Suppose further that for some point A

$$A \models \langle i \rangle [j] F$$

There is then B with $A \xrightarrow{i} B$ and $B \models [j]F$. Let C be such that $A \xrightarrow{k} C$. Using confluence we get D with $D \models F$. Furthermore $C \models \langle l \rangle F$. Since C was arbitrary, then $A \models [k]\langle l \rangle F$.

$\Leftarrow$: Suppose the axiom is true and we have the situation

The axiom is true for all interpretations. Let F be true in exactly those universes which are j-visible from B. Then $B \models [j]F$ and $A \models \langle i \rangle [j]F$. Using the axiom we get $A \models [k]\langle l \rangle F$ and $C \models \langle l \rangle F$. So there is a universe D -l-visible from C with F true — and we get

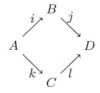

■

In ordinary modal logic we have as normal operators the usual one and also the identity operator. Let us give some applications of confluence there.

$\Box F \rightarrow \Diamond F$: We let j and l be the usual transition, and i and k be identity. Then confluence says that from A we can find D with $A \longrightarrow D$. This property is called seriality — we can always see new elements.

$\Box F \rightarrow F$: Let i, k and l be identity, while j is the usual transition. Confluence gives reflexivity.

$F \rightarrow \Box \Diamond F$: Let i and j be identity, while k and l are the usual transition. Confluence gives symmetry.

$\Diamond \Box F \rightarrow \Box F$: We let i, j and k be the usual transition, while l is identity. The axiom is called negative introspection — especially in the contrapositive form $\Diamond \neg F \rightarrow \Box \Diamond \neg F$. Confluence is often called Euclidean property.

7.2   Temporal logic

It is an old tradition to connect necessity with what is always true — no matter what the future bring. Here we can look at two temporal modalities — $[-]$ and $[+]$ — connected with past and future. The following is a straightforward description of a minimal temporal logic $\mathbf{K}_t$:

- We have two temporal modalities $[-]$ and $[+]$

- $[-]$ and $[+]$ satisfy

  - they are normal modalities
  - they are transitive — they satisfy
    * $[-]F \rightarrow [-][-]F$
    * $[+]F \rightarrow [+][+]F$
  - they are inverse to each other — they satisfy
    * $F \rightarrow [-]\langle+\rangle F$
    * $F \rightarrow [+]\langle-\rangle F$

A more expressive temporal logic has two operators

Since: $F\mathcal{S}G$ means "F is true since G is true"

$$b \models F\mathcal{S}G \Leftrightarrow \exists a \prec b.(a \models F \wedge \forall c.(a \prec c \prec b \rightarrow c \models G))$$

Until: $F\mathcal{U}G$ means "F is true until G is true"

$$a \models F\mathcal{U}G \Leftrightarrow \exists b \succ a.(b \models G \wedge \forall c.(a \prec c \prec b \rightarrow c \models F))$$

With the two new operators $\mathcal{S}$ and $\mathcal{U}$ we can define

$$
\begin{array}{rcl}
\langle+\rangle F & \Leftrightarrow & \mathsf{T}\mathcal{U}F \\
[+]F & \Leftrightarrow & \neg(\mathsf{T}\mathcal{U}\neg F) \\
\langle-\rangle F & \Leftrightarrow & F\mathcal{S}\mathsf{T} \\
[-]F & \Leftrightarrow & \neg(\neg F\mathcal{S}\mathsf{T})
\end{array}
$$

7.3 Description logic

The multi modal logic can be seen as a basic description logic. Let us say that we have a particular model of the multi modal logic. Then we view

Domain: The set of all universes

Attributes: The propositional variables which may or may not be true in a particular universe

Roles: The transitions between the universes

Over the domain we then have attributes as unary predicates and roles as binary predicates. Of course we could use fragments of first order logic instead of multi modal logic. The advantages of having the connections to multi modal logic is as follows

- First order logic with binary predicates is undecidable

- Multi modal logic is decidable

- To get a decidable calculus we must have restrictions on the binary predicates.

- The reasonable and natural restrictions are not so easy to find in first order logic but may be easier in multi modal logic

- The further development of description logic takes multi modal logic as a start, but then develops properties of especially the roles which helps expressivity of the language without hindering efficiency of the calculus

7.4   Epistemic logic

We have a multi modal logic where $[i]F$ is interpreted as "agent i knows F". We need two extra axioms for the knowledge operators

Positive introspection: $[i]F \to [i][i]F$

Negative introspection: $\neg[i]F \to [i]\neg[i]F$

Positive introspection is the same as our axiom **4** giving transitivity. We treat here some important examples of how we can use epistemic logic to describe situations.

Muddy children

- N children — each with its own epistemic operator

- 1 father who observes that at least one child has a muddy forehead

- each child can find out by observation which other children are muddy but not itself

- by asking the other children, each child is able to find out by reasoning and using the responses from the other children whether it is muddy or not

There is a protocol in N rounds where each child can find out whether it is muddy or not.

Round 1: Only a child which is the only muddy one, can and will answer. Then we have the answer for each of the children.

Round 2: The children now knows that at least 2 children are muddy. If one of them sees only one other with mud on the forehead, she and also the muddy one can conclude that they are the only muddy children. Then we have the answer for each of the children.

Round 3: The children now know that there are at least 3 muddy children. And then the round and further rounds go on.

Round N: The protocol may last until this round. Now the children will know that they are all muddy.

Common knowledge

To solve the muddy childrens problem we need iteration of the epistemic operators. We must not only know what we know and what other know, but also what other know about what we know and so on. Note the following distinctions

Shared knowledge: "Felles kunnskap" — what everybody knows

$$\mathcal{F}(G) \leftrightarrow \bigwedge_i [i]G$$

Common knowledge: "Allmenn kunnskap" — what everybody knows that every body knows and so on

$$\mathcal{A}(G) \leftrightarrow \bigwedge_i \bigwedge_j \cdots \bigwedge_k .[i][j] \cdots [k]G$$

Implicit knowledge: Logical consequences of what every agent
know

$$\mathcal{I}(G) \Leftrightarrow G \text{ is a consequence of all } [i]H$$

Coordinated attack

Two troops — one on each of two neighbouring hills — are ready
to attack the enemy on the plains below. We assume

- each troop must attack at the same time — then they win
  else they lose

- to coordinate the attack there is an agent running from
  one hill to the other

- the communication using the agent is unreliable — the
  agent may be lost or be caught by the enemy

We can show that there is no protocol which makes the communication absolutely reliable. To get it we need some common
knowledge between the two troops.

7.5 The Byzantine generals

- We have N generals and among them there are T traitors and the rest are honest

- Each general may decide to attack or not

- The generals can communicate reliably on a person-to-person basis — the communication is synchronous, we know when an expected message will arrive

- At the start each general has a preference whether to attack or not

- We want to find a protocol where in each round the generals communicate with each other about their preferences for attack — all generals follows the protocol, they answer when they should and the round closes when it should.

- There are no assumptions about truth and consistency in the reports by the traitors

- After a given a number of rounds the generals decide such that

  - all honest generals decide on the same action
  - if the honest generals have the same preference at the start of the rounds, then they also end up with the same preference

So the traitors may lie and tell different thing to different people. They try to cheat as best as they can. There is such a protocol if and only if $N > 3T$. So if there are not too many traitors we can get common knowledge among the honest generals. We shall give the main ideas behind this, but leave to the reader to get a detailed protocol. We treat the two main cases

- 1 traitor and 2 honest generals

- T traitors and more than $2T + 1$ honest generals

1 traitor and 2 honest generals

The three generals are

All communications go along the edges. Assume $\otimes$ is the traitor. The traitor behaves in a schizophrenic way — to $\odot$ he says that he is going to "attack", but to $\oplus$ he is saying that he is going to "not attack". There is no way that the honest generals can find out whether he or one of the other is the traitor.

1 traitor and 3 honest generals

The protocol is using two things

Majority choice: To find a preference for a general we ask many questions about it and use the majority rule to find what we think he prefers.

Common knowledge: For each general we not only asks directly his preference, but also what the other generals knows about his preference, and so on.

Now we have the four generals and their communications

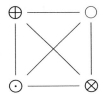

Say that the traitor is ⊗. Each general can then find out the preferences of the honest generals, but the information about the traitor is unreliable. The problem is that we do not know who is honest and who is the traitor. More precisely we have the following cases among the honest generals and the received messages

$3a + 0r$: Received either $4a + 0r$ or $3a + 1r$

$2a + 1r$: Received either $3a + 1r$ or $2a + 2r$

$1a + 2r$: Received either $2a + 2r$ or $1a + 3r$

$0a + 3r$: Received either $1a + 3r$ or $0a + 4r$

Most cases can be decided by a majority vote. The problem comes when an honest general receives $2a+2r$. Should he attack or retreat? Here we must use information about what the other generals convey indirectly.

Message tree

Each general receives messages about what the others generals are preferring. Say the generals are named 1 2 3 4. We organize the messages to each general in a **message tree** of the following form

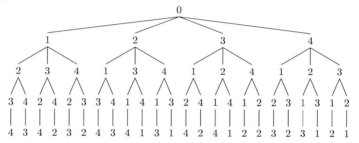

Each of the generals build such a tree. Each node have three information pieces

Node: Each node is given by a sequence of numbers. The down-most rightmost node is denoted by **04321**. If we have N generals, the message tree for each general has $N!$ many branches. Each branch starts with 0 and then use a permutation of 1 2 ... N.

Input: The message tree for each general is decorated by what the general hear about the preferences from the other generals. So for example node **0241** has as input what the general hears that general 1 says that general 4 says that general 2 has. The root node **0** is not given an input

Output: This is calculated from the inputs by a majority rule — starting from the leaf nodes and going upwards. The leaf nodes have output=input. Then for a node we look at the outputs from the sons and take the majority. If there is an even split we take as input *retreat*.

A general gets his calculated preference by

- Get the message tree for the general with nodes and input

- Calculate the output of the nodes in the message tree

- The output of the root is the preference for the general

In the calculation we used a default action if there was no majority at a node. It is essential that we have the same default throughout the calculation, but it does not matter whether it is *attack* or *retreat*.

The problem with this protocol is the computational complexity. The message tree may be enormous and it is not feasible to find the preferences in this way.

Using a global coin

By relaxing on some of the conditions and introducing a new device — a global coin — we can make a much better algorithm. The global coin is seen by all generals. After each round we toss the coin giving head or tail. Assume we have T traitors and $2T + 1$ honest generals. Each general remembers two pieces of information in each round

Preference: His preferred action — attack or retreat

Tally: The number of generals saying they have the same preference

Then for each honest general in each round there are two cases:

Tally $\geq 2T + 1$: Keep the preference to the next round.

Tally $\leq 2T$: Look at the global coin and get a preference for the next round — if head then prefer attack, else prefer retreat

We observe that in each round:

- Each honest general with tally $\geq 2T + 1$ have the same preference. Else there would be at least $T + 1$ honest generals preferring attack and $T + 1$ preferring retreat. Because of the traitors the tally may be less in the next round.

- There is a 50 % chance that the global coin will also show this preference and then in the next round all honest generals will have the same preference.

The protocol decides on the number of rounds N and when we stop there is only a 2^{-N} chance that we have not a good decision for the Byzantine generals. In each round we need $M(M-1)$ messages for M generals.

8

Games on finite arenas

8.1 Arena

We play 2-persons game on finite arenas and ask for winning strategies. The two players are called ∀belard and ∃loise. There is a finite and an infinite variant of the game. First the infinite parity game (IPG)

- The arena is built on a finite directed graph $\mathcal{G}$ and from each node there is at least one arrow out

- The nodes are partitioned into two sets — $V_\forall$ and $V_\exists$

- ∃loise chooses the arrow out for nodes $V_\exists$ — and ∀belard chooses the arrow out for nodes $V_\forall$

- Each node is assigned a natural number — called a level

- There is a starting node

- A run in a game is an infinite path through the nodes — starting with the starting node and following the arrows chosen by the two players.

- ∃loise wins a run if the maximal level met infinitely often is even

- ∀belard wins a run if the maximal level met infinitely often is odd

73

This is called the infinite parity game — IPG. There is also a finite parity game — FPG where the run lasts until we get to a loop (one node is visited twice). In FPG we look at the maximal level within the loop. If the level is even ∃loise wins, if it is odd ∀belard wins. We shall first develop a theory for the finite parity game and then show that it is generalized to the infinite parity game.

8.2 Game trees

Given a finite arena A. Above each node $u \in A$ we build a game tree – $\mathcal{G}_u$ – over u.

- The nodes in the game tree are nodes from the arena — and we have ∀-branchings and ∃-branchings.

- For wins in the FPG we look in each branch for the first place where a node is repeated. If the arena has N nodes, then we only have to look at the game tree up to height $N + 1$.

- For wins in the IPG we had to look at each branch — find the maximal level of the nodes which are repeated infinitely often. The branches are infinite.

In FPG we can decide whether ∃loise or ∀belard has a win starting from node u. The arena is divided into four parts

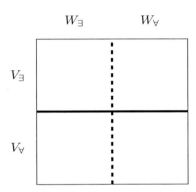

There are two partitions of the nodes

Who moves: We partitioned the nodes into $V_\exists$ and $V_\forall$

Who wins: The nodes where ∃loise wins — $W_\exists$. The nodes where ∀belard wins — $W_\forall$.

Over each point u in the arena we can construct a finite game tree $\mathcal{G}_u$. The game trees have $\forall$- and $\exists$-branchings. The downmost branching determines whether we are in $V_\forall$ or $V_\exists$. We need to look at all the branches to decide whether we are in $W_\forall$ or $W_\exists$ — here we need a PSPACE calculation.

The key to get a further analysis is to look at connections between points connected with edges. We have

Lemma 8.1 *For an* $\exists$-*node* u

$$u \in W_\exists \quad \Rightarrow \quad \exists v \leftarrow u.v \in W_\exists$$
$$u \in W_\forall \quad \Leftarrow \quad \forall v \leftarrow u.v \in W_\forall$$

and for an $\exists$-node u with a single edge from it we also get the converse

$$u \in W_\exists \quad \Leftarrow \quad \exists v \leftarrow u.v \in W_\exists$$
$$u \in W_\forall \quad \Rightarrow \quad \forall v \leftarrow u.v \in W_\forall$$

Similarly for an $\forall$-node.

Consider a point u with game tree $\mathcal{G}_u$ and a successor v to u connected with edge e and with game tree $\mathcal{G}_v$. The two game trees are intertwined. We picture this as

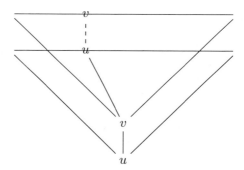

We reach the top in the game tree when we get to a loop in the branch. To get to the loop in the game tree over v we may have to extend the branches a little higher up to get to the

loop. This happens if the branch in the game over u ends with u. The loops in the branches of $\mathcal{G}_u$ which starts with edge e to v corresponds to the loops in the branches in $\mathcal{G}_v$.

Now to the proof of the lemma. Assume $u \in V_\exists \cap W_\exists$. Then there must be an edge $e : u \rightarrow v$ such that the branches in $\mathcal{G}_u$ starting with edge e shows that u is $\exists$-winning. These branches correspond to branches in $\mathcal{G}_v$ showing that v is $\exists$-winning.

For the converse argument we get this only when the edge e is the only edge starting from u.

The finite games are determined — either they are $\exists$-winning or they are $\forall$-winning. We get $w \in W_\exists \Leftrightarrow w \notin W_\forall$. From this we get

$$u \in W_\forall \Leftrightarrow u \notin W_\exists \Leftarrow \neg \exists v \leftarrow u.v \in W_\exists$$
$$\Leftrightarrow \forall v \leftarrow u.v \notin W_\exists \Leftrightarrow \forall v \leftarrow u.v \in W_\forall$$

And similarly for the rest of the lemma.

8.3   Choices

Given an arena some of the nodes may have choices — others have a single arrow leading out. The choices give rise to branchings in the game trees. In general the game tree is an AND-OR tree.

Choice node: A node with at least two edges going out

$\exists$-choice: A choice in the arena from an $\exists$-node

$\forall$-choice: A choice in the arena from an $\forall$-node

Subarena: Arena with the same number of nodes but fewer choices — some arrows are deleted.

∃-positional strategy: Subarena where we have removed all ∃-choices and kept all ∀-choices.

∀-positional strategy: Similar.

We now want to reduce the number of choices in an arena while keeping the four parts of the arena intact.

Theorem 8.2 *Given a finite arena $\mathcal{A}$. We can find an ∃-positional strategy $\mathcal{A}^{\exists}$ and an ∀-positional strategy without changing the winning nodes for any of the strategies.*

- The positional strategies $\mathcal{A}^{\exists}$ and $\mathcal{A}^{\forall}$ can be considered as strategies for the players — $\mathcal{A}^{\exists}$ for ∃loise and $\mathcal{A}^{\forall}$ for ∀belard.

- If ∃loise has a win from node u, then she has a strategy where her moves depends only on the node she is in. The game tree over $u \in W_{\exists}$ contains only ∀-branchings.

- In general the game trees are AND-OR trees, but with the positional strategies they are AND-trees. This means that all nodes in the game tree over the node $u \in V_{\exists} \cap W_{\exists}$ can be assumed to be ∃-wins.

The proof is by induction over the number of choices. The induction is used to prove that the subarenas have the desirable properties, but the construction of the subarenas is straightforward. To get rid of the ∃-choices we do the following in order

- We get rid of the $\exists$-choices in $V_\exists \cap W_\exists$ for each node $u \in V_\exists \cap W_\exists$ do: There must be an edge $e : v \leftharpoonup u$ such that all branches following e through v are $\exists$-winning. Keep this edge and remove all the other edges from u.

- We now observe that there are no edges from $W_\forall$ to $W_\exists$

- Remove the remaining $\exists$-choices from the arena in any way to get an arena without $\exists$-choices.

We have a similar construction to get rid of the $\forall$-choices. In the proof below we prove by induction that we can do this pruning without changing the winning or the losing nodes of the arena.

Induction start

With no choices in arena $\mathcal{A}$ we obviously have the theorem.

Induction step — remove choices in $V_\exists \cap W_\exists$

- Assume we have a choice in $u \in V_\exists \cap W_\exists$ and we have edge $e : u \rightharpoonup v$ where all branches from u which follows e to v are $\exists$-winning.

- Remove the other edges from u to get subarenas $\mathcal{A}^\star$ with fewer choices. In $\mathcal{A}^\star$ we have an $\exists$-win from u.

- By induction we get subarenas $\mathcal{A}^\exists$ and $\mathcal{A}^\forall$ of $\mathcal{A}^\star$ which are positional strategies for $\mathcal{A}^\star$.

- In the change from $\mathcal{A}$ to $\mathcal{A}^\star$ we remove some $\exists$-choices and keep all $\forall$-choices. So we have $W_\forall \subseteq W_\forall^\star$ and we must prove that we have equality. Assume we have $x \in W_\forall^\star \cap W_\exists$. Then $\forall$belard can force a path from x to u. This means that $u \in W_\forall^\star$ and we have a contradiction.

Induction step — remove choices in $V_\forall \cap W_\forall$

Similar to the above.

Induction step — No edges between $W_\exists$ and $W_\forall$

We assume that we have removed the choices in $V_\exists \cap W_\exists$ and $V_\forall \cap W_\forall$. Suppose that we have an edge e from $v \in W_\forall$ to $u \in W_\exists$. Then

- There must be a choice in v — another edge d starting from v. Else we would have $v \in W_\exists$. Since we have removed the choices in $V_\forall \cap W_\forall$ we must have $v \in V_\exists$.

- Let $\mathcal{A}^\star$ be the subarena where we have removed the edge e. There are fewer $\exists$-choices and $W_\forall \subseteq W_\forall^\star$. By induction we get positional strategies $\mathcal{A}^\exists$ and $\mathcal{A}^\forall$ of $\mathcal{A}^\star$.

- $\forall$belard can force a win from v even if $\exists$loise starts by choosing edge e to node u. But then $\forall$belard can force a path from u and back to v. This path is within $\mathcal{A}^\star$ and can be continued within $\mathcal{A}^\star$ and gives an $\forall$-win. This means that $u \in W_\forall^\star$.

- On the other hand starting from u ∃loise can force a win as long as she uses edge e. So she can force a path to v. In v she can choose edge d instead and will get a win in $\mathcal{A}$ since $v \in W_\exists$. But this win is contained within the subarena $\mathcal{A}^\star$ and we get $u \in W_\exists^\star$ and a contradiction.

We have a similar contradiction if there are edge going the other way — from $W_\exists$ to $W_\forall$.

Induction step — Removing extra edges

We assume that we have already done the pruning above. There still may be extra choices. It does not matter which we prune. After the pruning if we start within $W_\exists$ the whole game will be within it, and similarly if we start within $W_\forall$.

8.4 Infinite positional strategies

The positional strategies above are the key to connecting the finite games with the infinite game. Say that we have removed all ∃-choices. Then a branch in the game tree will only contain ∀-branchings. This means that it is up to ∀belard to choose when he meets a loop whether to enter the loop or continue without entering it. So if ∀belard wants to meet a special node in a loop he can choose the run so that this loop is the first he enters.

Within a finite arena we get positional strategies for FPG. Interestingly every positional strategy for FPG is also a positional strategy for IPG — and conversely.

From FPG to IPG: Assume we have an $\exists$-positional strategy which loses in IPG. Then there is a branch in the game tree where the maximum level of an infinitely repeated node — say u — is odd. In the game tree we can find another branch where the node u is the first repeated node. The positional strategy loses also in FPG.

From IPG to FPG: Assume we have an $\exists$-positional strategy which loses in FPG. There is an $\exists$-losing loop in a branch in the game tree. Then $\forall$belard can force to stay within that loop forever and the strategy is also losing in IPG.

So we have the following PSPACE procedure to find winning nodes in the infinite parity game

- Given an IPG

- Using the FPG we partition the nodes into winning nodes for $\exists$loise and $\forall$belard.

- Given a winning node in FPG for $\exists$loise, we can find a positional strategy for her.

- This strategy is also a winning strategy for IPG.

- The arguments generalizes to other winning conditions than the parity property of loops used here.

9

Decision problems

9.1 Games using input

In the games above we had only interaction from the two players. We extend the games to have input and various forms of indeterminacy

- there is a finite input alphabet $\mathcal{A}$

- as input we have the full binary tree decorated with symbols from $\mathcal{A}$

- the two players have different uses of the input

- at the start we begin with the symbol at the root

- ∃loise reads a symbol from the alphabet — symbol reader

- ∀belard chooses whether the next symbol is in the tree to the left or to the right — path finder

- the arrows out from $V_\exists$ are each decorated with a symbol from the alphabet — and ∃loise must choose an arrow with the symbol she reads

- at the same time as ∀belard chooses an arrow he chooses also the direction — left or right — where ∃loise shall read her next symbol

- ∃loise or ∀belard can guess an indeterminacy

We get the game without input if the input alphabet consists of only one symbol. We can also extend the game to a game with more than one input by letting ∃loise read more symbols and let ∀belard have more choices of directions. This is straightforward and can be defined in many equivalent ways.

Using the constructions from the games with finite arenas we can decide which player has a winning strategy for the finite or the infinite parity game.

9.2   Second order monadic theory — S2S

As universe we have the binary numbers

- Two unary functions

  - $x \mapsto x0$
  - $x \mapsto x1$

On this universe we have second order monadic logic.

- first order quantification over numbers

- second order quantification over monadic predicates

- $s = t$ — equality between numbers

- Xs — the number s satisfies X

Within the logic we may define

ε	$: \neg\exists y(y0 = \varepsilon \vee y1 = \varepsilon)$	— empty number
$a < b$	$: \forall X(Xa \wedge \forall y(Xy \rightarrow Xy0 \wedge Xy1) \rightarrow Xb)$	— precedes
$C(X)$	$: \forall x.\forall y(Xx \wedge x < y \rightarrow Xy)$	— chain
$IC(X)$	$: C(X) \wedge \forall y.\exists z.(Cy \wedge y < z \rightarrow Cz)$	— infinite chain
$FC(X)$	$: C(X) \wedge \neg\forall y.\exists z.(Cy \wedge y < z \rightarrow Cz)$	— finite chain
$F(X)$	$: \forall y.\forall Z.(Xy \wedge Zy \wedge C(Z) \rightarrow FC(Z))$	— finite set

S2S describes IPG

Given an IPG we can use S2S to describe it in such a way that a win in IPG correspond to validity of a sentence in S2S. This is done as follows

Finite run: Binary number

Infinite run: Infinite chain of binary numbers, or a branch in the full binary tree

States: Monadic predicate with argument a binary number

Arena: A finite number of monadic predicates connected to each other with transitions described by sentences in S2S.

And so on. We leave the actual description to the reader.

IPG decides validity in S2S

Conversely we can use games to simulate second order logic.

The second order monadic logic is built up using

- $s \mapsto s0$

- $s \mapsto s1$

- $s = t$

- Xs

We simplify things by getting rid of the first order constructions by using

- $SING(X)$ — X is a singleton

- $X \subseteq Y$ — subset

- $X \subseteq_0 Y$ — subset after moving one to the left

- $X \subseteq_1 Y$ — subset after moving one to the right

We use IPG with inputs. Given a formula $F(X, \ldots, Z)$ in S2S we define an IPG where we have inputs corresponding to the free variables $X, \ldots, Z$ and a win if the formula is true. This is done by the build up of formulas:

Atomic formulas: Trivial

Conjunction: Use pairs of arenas

Negation: Change ∃-nodes to ∀-nodes and conversely

Quantifier: We treat second order ∃-quantifier using indeterminacy

Note the construction for the second order quantifier. The free variables are treated as inputs in the game. In logic we can hide a variable using a quantifier. In the game we can hide it treating it as an extra indeterminacy.

The result is an IPG with a win exactly when the sentence in S2S is valid.

9.3 Decidability of arithmetic

As a datastructure the unary numbers have some deficiencies. When Skolem introduced it in 1923 he had to include the primitive recursive functions to have expressive power. Gödel in 1930 used addition and multiplication. To express syntax within such a framework he had to use extra number theoretic arguments — he used the Chinese remainder theorem. Even then syntax could only be expressed with Σ_1-statements and not Δ_0-statements as we do within the datastructure of binary trees. In this section we consider some of these phenomena.

Arithmetic with addition is decidable

As language we have

- Constants — 0 1

- Ternary relation — $x + y = z$

We can then express with predicate logic successor, less than and much more. Presburger and Skolem proved that we can decide the truth values of sentences in predicate logic over this language. Given a formula $F(x, y, \ldots, z)$ in the language we build an automaton $\mathcal{A}$ with inputs for each free variable in F such that $F(k, l, \ldots, m)$ is true if and only if the automaton accepts the inputs corresponding to $k, l, \ldots, m$.

Numbers: Binary numbers read from right to left

$x + y = z$**:** Basic automaton for adding binary numbers

$\wedge$ **and** $\vee$**:** Intersection and union of automata

$\neg$**:** Using subset construction to get a deterministic automaton and then take complement

$\forall$ **and** $\exists$**:** Using conjunctive and disjunctive indeterminacy

The binary numbers are not unique — we may have a number of 0's on the left. We make them unique by having infinite streams as input by adding on extra 0's. The result is an automaton that can decide the truth values of sentences in our language.

Arithmetic with multiplication is decidable

Skolem published a paper on this after having learned about Presburgers publication. We make some changes in the above argument

- The automata takes finite trees as input

- Numbers are represented using prime number representation with binary numbers as exponents to the prime numbers

- Multiplication is reduced to addition of binary numbers — to each prime number we add the exponents to them

We get the decidability of arithmetic with only multiplication.

Arithmetic with addition and multiplication may need extra quantifier

But we can still not represent syntax. We get into problems where we represent properties like "being a tail element" which we treated above in the setting of pairs. For Gödel the way around was to use the Chinese remainder theorem.

Let us start with a representation of all pairs $\langle x, y \rangle$ of natural numbers with $x < 3$ and $y < 5$. Since there are 15 such pairs we need at least 15 distict numbers to represent them. It turns out that the numbers from 0 to 14 suffice. This can be seen from the table on the next page. As an example we let the number 8 represent the pair $\langle 2, 3 \rangle$. From number theory we know that it is essential that the numbers 3 and 5 are relatively prime. This observation generalizes to arbitrary long sequences.

Number	Remainder modulo 3	Remainder modulo 5
0	0	0
1	1	1
2	2	2
3	0	3
4	1	4
5	2	0
6	0	1
7	1	2
8	2	3
9	0	4
10	1	0
11	2	1
12	0	2
13	1	3
14	2	4

Theorem 9.1 (Chinese remainder theorem) *Suppose that we have numbers $d_0, \ldots, d_{n-1}$ which are relatively prime. For a number x let r_i be the remainder of x modulo d_i. We then let x code the finite sequence $\langle r_0, \ldots, r_{n-1} \rangle$. Then the coding function*

$$x \mapsto \langle r_0, \ldots, r_{n-1} \rangle$$

from numbers $< d_0 \times \cdots \times d_{n-1}$ to sequences in $d_0 \times \cdots \times d_{n-1}$ is bijective (i.e. 1-1 and onto).

Proof. Let us first show that the function is 1-1. Assume we have two numbers $x \leq y < d_0 \times \cdots \times d_{n-1}$ with the same code

$\langle r_0, \ldots, r_{n-1} \rangle$. But then all d_i will divide the difference $y - x$. Since the d_i are relatively prime then the product $d_0 \times \cdots \times d_{n-1}$ will also divide $y - x$. But $0 \leq y - x < d_0 \times \cdots \times d_{n-1}$ which shows that $0 = y - x$ and hence $x = y$.

So the coding function is 1-1. But both the domain and the range of the function are finite and contain the same number of elements. Therefore the coding function is also onto. ∎

We can therefore code long sequences of numbers if we have long lists of relatively prime numbers. We now use

Theorem 9.2 *Let* $n > 0$ *and* $d = n! =$ *the factorial of* n. *Then the numbers* $1 + d, 1 + 2d, \ldots, 1 + (n+1)d$ *are relatively prime.*

Proof. Consider $1 + id$ and $1 + jd$ with $1 \leq i \leq j \leq n+1$. Let p be a prime number dividing both numbers. Then $p > n$ and p divides the difference $(j-i)d$. Since p is a prime $> n$, then p cannot divide $d = n!$. Hence p divides $j - i$. But $0 \leq j - i < n$ and hence $i = j$. ∎

This gives us the famous β-function of Gödel.

$$\beta(c, d, i) = \text{the remainder of } c \text{ modulo } 1 + (i+1)d$$

Theorem 9.3 *Let* $a_0, \ldots, a_{n-1}$ *be a sequence of natural numbers. Then there are* c *and* d *such that for all* $i = 0, 1, \ldots, n-1$ *we have*

$$\beta(c, d, i) = a_i$$

So we get a Σ_1-formula representing that something is code for a sequence. We need the extra quantifier to get large enough numbers to feed into the β-function. We used factorials in getting large enough numbers and hence went beyond the addition and multiplication used in the language. These large numbers are hidden by the extra $\exists$-quantifiers.

But for the language of pairs it is much simpler.

9.4 Processes

Using the methods of games with a finite arena we can decide properties of processes. We think of a process as running through a finite number of states depending on input from the user or from the environment. We assume that we have only a finite number of states — and thereby exclude cases where the states may contain an arbitrary element of an infinite datastructure. We can then use appropriate assignments of levels to the states to decide

Reachability: Using the finite game we can find out whether a certain state is reached or not.

Liveness: Using the infinite game we can decide whether we can from any state reach a given starting state — whether we can always boot the process.

An essay on logic

Thinking from assumptions

In logic we think from assumptions. An extreme case is when the assumptions are simply wrong — as in proofs by contradiction. Say we want to prove that there are infinitely many prime numbers. This is done as follows

- Assume there are just a finite number of prime numbers — $p_0, p_1, \ldots p_k$

- Multiply the prime numbers together and add 1 to get — $P = p_0 \cdot p_1 \cdots p_k + 1$

- The number P has a prime number factorization, but none of the prime numbers $p_0, p_1, \ldots, p_k$ divides P

- Contradiction and the assumption is wrong.

This thinking from assumptions is the key to understanding logic — and other formal disciplines. We use it all the time. Note some of the typical assumptions

- something is finite

- something is computable

- something is valid

- something is decidable

None of these assumptions are decidable. The point is to show that from for example assuming something is finite, we get to know that something else is finite. The assumptions about finiteness propagates through the argument in some way. As logicians we investigate how we can make such assumptions propagate. We are sort of like magicians — they do not produce the rabbits from nothing, but they introduce them from the beginning and let them propagate in — for the audience — a hidden way through the performance.

The four levels of reasoning

Following Frege we have four levels where our reasoning is done

Physical level: This is the level where we have white, black and grey smudges on the paper which indicates signs and symbols. There could also be auditory signals and smells and touches.

Syntactical level: We have interpreted the physical signals as letters and numbers. We have passed from the physical tokens to the syntactical types. Now we can decide whether something occurs twice.

Semantical level: The syntax is given an interpretation — we know for example the truth conditions.

Pragmatic level: We use the semantic signs to act within our world. We use semantic sentences to judge / to question / to embarass and so on.

There may or may not be a barrier between these levels. We do not know and for the development of logic we do not care. But we act as if there where such a barrier. So let us say something about the three barriers

Syntactic barrier: Animals seem to be able to bridge this barrier, humans do it all the time and machines are pretty good in doing it. But we develop and choose environments where this barrier can be bridged, or where we can assume that it is. This is an important effect of the civilization process.

Semantic barrier: There are a number of discussions whether this is a real barrier or not. We can think of the computer as syntax machines, and then we have the discussions about strong or weak AI as discussions about the existence of such a barrier. But for practical purposes it is. We make computers perform semantic work by observing that the semantics of the input is propagated through the computation and ends as a semantic of the output. This is what we do with electronic calculators and all kind of computers — at least so far.

Pragmatic barrier: Even if we understand a sentence there are a number of actions it can be used to. How to do things with words is a big topic and there are no reason that we should dismiss it as an unimportant thing and that it is only something we can do automatically and do not need to reason about.